THE FOOD HYGIENE HANDBOOK

Richard A. Sprenger B.Sc., D.M.S., F.I.E.H.

Director of Environmental Services
Doncaster Metropolitan Borough Council

The Institution of Environmental Health Officers

First Published 1982
Second Edition 1985
Third Edition 1986
Fourth Edition 1987
Fifth Edition 1989

"Vue Pointe", Spinney Hill, Sprotbrough,
Doncaster DN5 7LY
Tel: Doncaster (0302) 850007

ISBN 0 9508561 7 7

Published for
The Institution of Environmental Health Officers,
Chadwick House,
Rushworth Street,
London SE1 0QT
Telephone: 01 928 6006

Printed by Garnett Dickinson Print Limited, Rotherham and London

Contents

Introduction

The incidence of food poisoning remains at an unacceptably high level. During 1988 approximately 25,000 cases were reported and this is only the tip of the iceberg, as many people do not bother to report mild attacks of food-borne illness.

The cost of food poisoning can be high, not only to the country through working days lost, but also to employers and employees engaged in food handling. The owners of premises implicated in food poisoning will lose business and employees may lose their jobs. Food handlers who contravene food legislation may be prosecuted and an outbreak of food poisoning could cost firms responsible thousands of pounds in damage claims. Negligent food handlers may poison themselves or even worse be responsible for the death of a person consuming food which they have prepared.

Rigorous enforcement of existing food hygiene legislation is important but is not, in itself, sufficient to prevent food poisoning. Food poisoning is normally caused by negligence or ignorance and consequently most experts in food hygiene believe that a reduction in the current depressing statistics can only be achieved by the education of food handlers. One mistake by an untrained food handler in the most modern food premises can cause a serious outbreak of food poisoning.

To be effective, principles of food hygiene should be taught systematically as an important part of early training. Hygiene should become a way of life for all food handlers as soon as they embark on their careers as, once ingrained, bad habits are most difficult to change.

This booklet is intended to supplement the Basic Food Hygiene Course of the Institution of Environmental Health Officers and provided that persons engaged in food handling operate in accordance with the principles laid down, they will not be responsible for a food poisoning outbreak.

After reading the book you should:

(1) know the causes of food poisoning;
(2) know how to prevent food poisoning;
(3) be aware of the standards of personal hygiene required by food handlers;
(4) know how to use refrigerators and freezers;
(5) understand stock rotation;
(6) know how to dispose of waste food without causing problems;
(7) know the common pests found in food premises and how they can be controlled;
(8) understand cleaning procedures;
(9) understand how food legislation affects you.

Food hygiene

Defining food hygiene

Food hygiene is more than just cleanliness; it includes all practices involved in:

(1) protecting food from risk of contamination, including harmful bacteria, poisons and foreign bodies;
(2) preventing any bacteria present multiplying to an extent which would result in the illness of consumers or result in early spoilage of the food;
(3) destroying any harmful bacteria in the food by thorough cooking or processing.

The cost of poor hygiene:

(1) food poisoning outbreaks and sometimes death;
(2) food contamination and food complaints;
(3) pest infestations;
(4) waste food due to spoilage;
(5) the closure of food premises by local authority action;
(6) fines and costs of legal action taken because of contraventions in hygiene legislation, or because of the sale of unfit or unsatisfactory food;
(7) civil action taken by food poisoning sufferers;
(8) loss of production.

All of these factors will contribute to a lowering of profits. If the commercial viability of the premises is threatened, employees may lose overtime, bonuses or even their jobs. It is therefore in the best interests of everyone involved in the preparation and handling of food to observe the highest standards of food hygiene.

The benefits of good hygiene:

(1) satisfied customers, a good reputation and increased business;
(2) compliance with the law;
(3) increased shelf-life of food;
(4) good working conditions, higher staff morale and lower staff turnover.

All of these factors will contribute to higher profits.

Food poisoning and food-borne diseases

Food poisoning

Food poisoning is an unpleasant illness which usually occurs within 1 to 36 hours of eating contaminated or poisonous food. Symptoms normally last from 1 to 7 days and include one or more of the following: abdominal pain, diarrhoea, vomiting and nausea.

Food poisoning may be caused by:

(1) bacteria or their poisons;
(2) viruses;
(3) chemicals such as insecticides and weedkillers;
(4) metals such as lead, copper and mercury;
(5) poisonous plants such as deadly nightshade and toadstools.

Bacterial food poisoning is by far the commonest and in some instances may result in death.

Common food poisoning bacteria	% of reported cases for which they are responsible
Salmonella	80% to 90%
Clostridium perfringens	5% to 15%
Staphylococcus aureus	1% to 4%

Food-borne diseases

Other bacterial illnesses which can be transmitted via food include typhoid, paratyphoid, tuberculosis, dysentery and brucellosis. However, unlike food poisoning, only small numbers of bacteria are required to cause the illness; a multiplication of bacteria within the food is not necessary.

The bacteria responsible are found, among other places, in man's intestines and the chain of infection is the same as that involved in food poisoning, i.e.

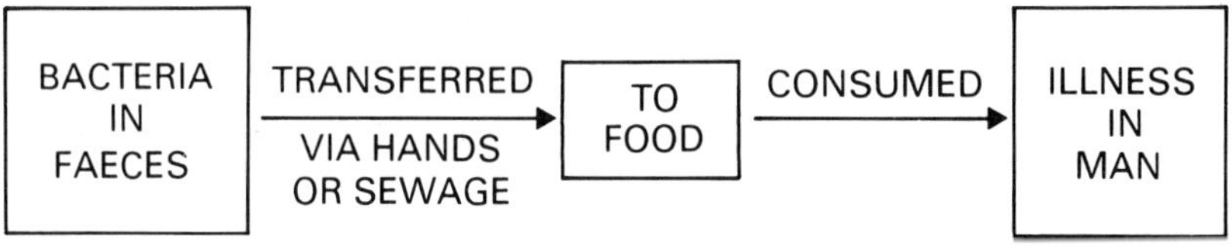

High-risk foods

High-risk foods are usually considered as those which support the multiplication of harmful bacteria and are intended for consumption without treatment such as cooking, which would destroy such organisms. These foods are usually proteins and require refrigerated storage. They must always be kept apart from raw foods. Examples include:

(1) all cooked meat and poultry;
(2) cooked meat products including gravy and stock;
(3) milk, cream, artificial cream, custards and dairy produce;
(4) cooked eggs and products made from eggs, for example, mayonnaise;
(5) shellfish and other seafoods;
(6) cooked rice.

All of the above are frequently implicated in outbreaks of food poisoning, especially poultry and cooked meat. Unfortunately contaminated food usually looks, tastes and smells completely normal and cannot be detected without laboratory examination.

The incidence of food poisoning

THE ANNUAL NUMBER OF CASES OF BACTERIAL FOOD POISONING IN ENGLAND AND WALES (1978-1988)

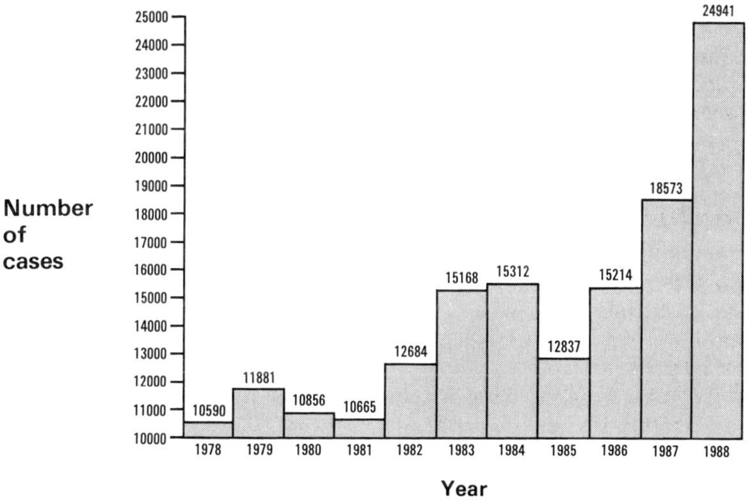

(P.H.L.S. Communicable Disease Surveillance Centre, Unpublished)

There are many more unreported incidents, despite the fact that it is a legal requirement to notify cases and suspected cases of food poisoning to local authorities for investigation.

Bacteria

Bacteria are microscopic organisms, often referred to as germs, which are found everywhere, including on and in man, on food, in water, soil and air.

Most bacteria are harmless and some are essential, for example, for breaking down decaying matter, or in cheese and yogurt manufacture. However, a small number of bacteria cause food spoilage and some, known as pathogens, are responsible for causing illness.

The number of bacteria present in food may be used to determine whether or not the food has been handled correctly.

Requirements for bacterial growth

Bacteria responsible for causing food poisoning need the following conditions to enable them to grow.

Warmth

The best temperature for the growth of food poisoning bacteria is 37°C (body temperature), although they can grow quite quickly between 20°C and 50°C. To prevent their growth we must ensure that the temperature of food is kept below 10°C or above 63°C. The temperature range of 10°C to 63°C is often referred to as the **"danger zone".** Some bacteria are able to produce spores which enable them to survive adverse conditions such as high temperatures.

Food and moisture

High protein foods are preferred especially, meat, poultry and dairy produce. Foods such as dried egg or milk powder do not provide the conditions necessary for growth. However, once water or milk is added to the powder, any bacteria present will start growing. It is essential, therefore, to use such food as soon as possible after adding water.

Other foods which do not support bacterial growth are those containing high concentrations of sugar, salt, acid or other preservatives.

Time

Given the right conditions of food, moisture and warmth, some bacteria can divide into two every 10 minutes. This process is known as binary fission. If there is sufficient time, a few bacteria can multiply to such an extent that there are enough present to cause food-poisoning. For this reason it is essential that high-risk foods are not left in the **danger zone** for longer than is absolutely necessary.

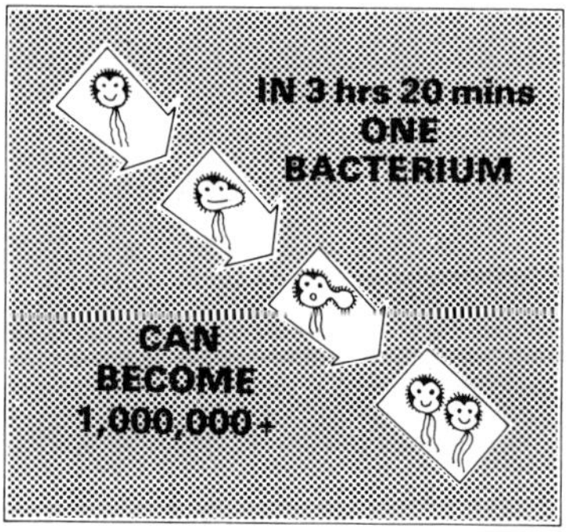

Food contamination

To prevent the consumption of unsound and unsafe food it is essential that contamination of high-risk food is kept to a minimum.

There are two types of contamination of high-risk food:

(1) *bacterial contamination* – which usually occurs within food premises because of ignorance, inadequate space, poor design or because of food handlers taking short cuts. Contamination of this sort is the most serious and may result in food spoilage, food poisoning or even death;
(2) *physical contamination* – by foreign bodies which may be dangerous, for example, glass or nails but is normally unpleasant and a nuisance.

Sources of food poisoning bacteria

(1) The person

People commonly harbour food poisoning bacteria in the nose, mouth, intestine and also on the skin. Food may be contaminated directly by the hands, sneezing or coughing, or indirectly by sewage contaminated water. All water used in food premises should be suitably treated, for example, by chlorination. The Aberdeen typhoid outbreak was due to cans of Argentinian corned beef which had been cooled in sewage-polluted water.

(2) Raw food

Raw food is particularly hazardous especially, red meat, poultry (up to 80% of frozen birds may carry salmonella), untreated milk, eggs and shellfish such as oysters. Raw food should always be kept separate from high-risk food. The liquid from defrosting foods, especially frozen poultry, must not be allowed to contaminate wiping cloths, high-risk food or equipment used for high-risk food.

Soil harbours harmful bacteria, and care must be taken when bringing vegetables into food rooms.

(3) Insects

Several insects may transmit food poisoning bacteria to food. Flies and cockroaches present the greatest hazard because of their feeding habits and the sites which they visit. Flies often land on animal faeces where they pick up large numbers of bacteria on their hairy bodies. In addition they defecate and vomit previous meals back onto the food as they feed.

Cockroaches often live in sewers and commonly feed on infected waste. They hide in the most inaccessible places in food rooms and may carry food poisoning organisms on their legs and bodies to food and equipment on which they walk.

(4) Rodents

Both rats and mice commonly excrete organisms such as salmonellae. Contamination of food may occur from droppings, urine, hairs and gnawing. Food-contact surfaces on which rodents have walked must be sterilized before use. Food suspected of being contaminated by rodents must be destroyed.

(5) Animals and birds

Both domestic and wild animals are known to carry harmful bacteria on their bodies and in their intestines. Furthermore, dirt can be transferred to food from their feet, and hairs and feathers may end up in the food. Pets must always be kept out of food rooms. Terrapins are occasionally implicated in cases of food poisoning through contact with infected water. Other incidents have occurred because of contamination of food by bird droppings.

(6) Dust

There are always large numbers of bacteria in dust and floating about in the air. Open food should always be covered when cleaning is carried out, especially dusting and sweeping.

(7) Refuse and waste food

Waste and unfit food must not be allowed to accumulate in food rooms. Care must be taken to avoid contamination of food from waste either directly or indirectly. Food handlers must wash their hands after placing refuse in dustbins.

Sources of food poisoning bacteria.

Vehicles and routes of bacterial contamination

Sometimes harmful bacteria pass directly from the source to high-risk food, but, as bacteria are usually static and the sources may not be in direct contact with food, the bacteria rely on other things to transfer them to food. These things are known as **vehicles** and the main ones are:

(1) hands; (2) cloths and equipment;
(3) hand-contact surfaces; (4) food-contact surfaces.

Indirect contamination using an intermediate vehicle is by far the commonest, for example, the passage of bacteria from the intestine of a food handler to food via the hands, after using the toilet. Where contamination is passed from raw food to high-risk food via, for example, a worktop, this is known as **cross-contamination.** The path which bacteria use to transfer from the source to the food is known as the **route.**

SOURCES, VEHICLES AND ROUTES OF CONTAMINATION

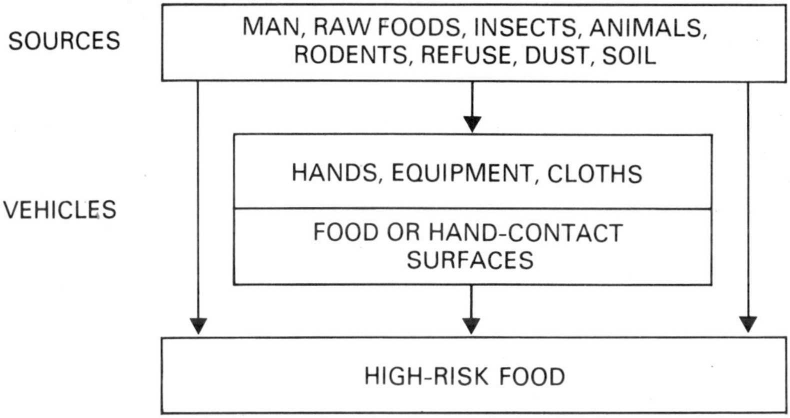

Physical contamination

Foreign bodies found in food may be brought into food premises with the raw materials or introduced during storage, preparation, service or display.

Although there are many different types of foreign bodies, it is essential that managers are aware of those commonly found in their particular sector of the food industry and that they exercise all due diligence to secure their removal or prevent their introduction. Food handlers must observe all company rules and take appropriate precautions to ensure that they are not responsible for the contamination of food.

Foreign bodies which often result in food complaints include:

(1) bolts, nuts, wire, staples and other pieces of metal – often found after maintenance and repair work;
(2) cardboard, string and polythene – often introduced from packaging;
(3) rodents, rodent hairs, feathers and droppings;
(4) sweet papers and cigarette ends;
(5) items from personnel such as earrings, fingernails, hair, buttons and pentops;
(6) glass;
(7) cleaning materials;
(8) mould;
(9) wood splinters;
(10) grease and oil;
(11) flaking paint or rust.

Essential points to remember about food poisoning

(1) Thousands of food poisoning incidents occur each year; they result in loss of business and jobs as well as the deaths of customers.
(2) Food which causes food poisoning looks, tastes and smells normal.
(3) The main reason for food poisoning is storage of high-risk foods at room temperature.
(4) Food poisoning bacteria are everywhere.
(5) Given the right conditions, warmth, food/moisture and time, bacteria will multiply rapidly.

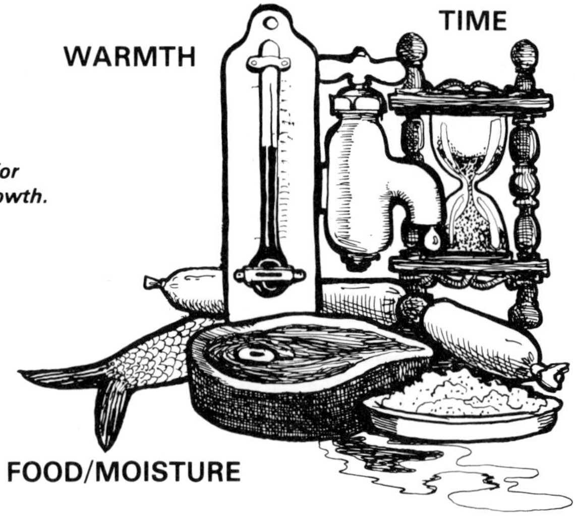

Conditions for bacterial growth.

THE TEN MAIN REASONS FOR FOOD POISONING

1. Food prepared too far in advance and stored at room temperature, i.e. not under refrigeration
2. Cooling food too slowly prior to refrigeration
3. Not reheating food to high enough temperatures to destroy food poisoning bacteria
4. The use of cooked food contaminated with food poisoning bacteria
5. Undercooking
6. Not thawing frozen poultry for sufficient time
7. Cross-contamination from raw food to cooked food
8. Storing hot food below 63°C
9. Infected food handlers
10. Use of leftovers

By courtesy of Dr. D. Roberts,
Food Hygiene Laboratory, Central Public Health Laboratory, Colindale.

The prevention of food poisoning

In most cases of food poisoning a chain of events takes place and, if we are to reduce the incidence of illness, this chain must be broken.

The food poisoning chain.

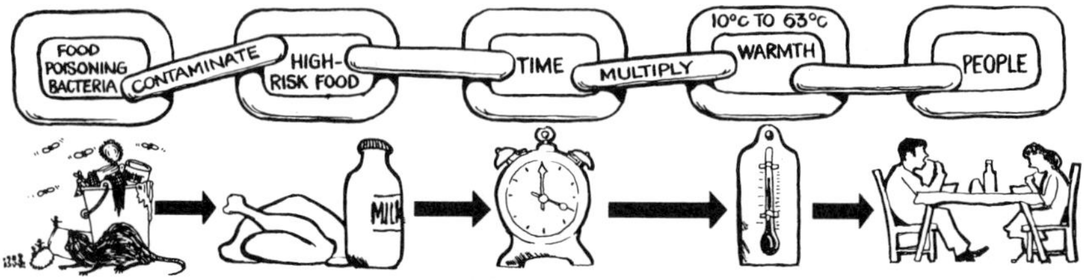

There are three main ways of breaking the food poisoning chain:

(1) protecting food from contamination;
(2) preventing any bacteria within food from multiplying;
(3) destroying those bacteria present within the food.

Protecting food from contamination by:

(1) keeping food covered wherever possible;
(2) not using unsuitable, defective or dirty equipment;
(3) not using dirty wiping cloths. Disposable cloths are preferable;
(4) only handling food when unavoidable. Tongs, plates and trays should be used in preference to hands;

Avoid handling food.

(5) separating raw and cooked food at all stages of preparation, storage and distribution. The same equipment and working surface must not be used to handle raw and high-risk foods;
(6) preventing insects, animals and birds from entering food rooms or coming into contact with food;
(7) storing food in rodent-proof containers and ensuring that the lids are tightly replaced after use;

(8) maintaining the highest standards of personal hygiene at all times;
(9) all food handlers wearing suitable protective clothing;
(10) not handling parts of crockery or cutlery that come into contact with food, for example, knife blades or inside glasses and cups;
(11) removing unfit or waste food and refuse promptly and keeping them apart from high-risk food;
(12) keeping food and equipment off the floor;
(13) ensuring that the liquid from thawed frozen meat and poultry does not come into contact with high-risk food or surfaces and equipment used for high-risk food;
(14) using the correct cleaning procedures;
(15) not using wash-hand basins for washing food or food equipment and not using food sinks for hand washing.

Food should not be washed in facilities provided for personal hygiene.

Preventing any bacteria within food from multiplying by:

(1) storing food out of the **danger zone.** Food should be kept below 10°C, for example, in a refrigerator, or kept above 63°C, for example, in a bain marie;
(2) ensuring that during preparation, food is within the **danger zone** for as short a time as possible. High-risk food must not be left in the ambient temperatures of kitchens or serving areas, unless in the course of preparation or needed for immediate consumption;
(3) using suitable preservatives such as salt and sugar;
(4) not allowing dried foods to absorb moisture.

Destroying those bacteria within food by:

(1) thorough cooking;
(2) heat processing such as pasteurization, sterilization or canning.

A combination of a suitable temperature and sufficient time is always required to destroy bacteria. The time and the temperature required will depend on the particular organism. For example, spores of *Clostridium perfringens* are much more heat resistant than salmonella bacteria.

Personal hygiene

Most people carry some type of food poisoning organism at one time or another, and food handlers have a moral and legal responsibility to observe high standards of personal cleanliness to ensure that they do not contaminate food.

Hands and skin

As the hands are in direct contact with food, they are the main route for transferring food poisoning bacteria. Hands must be kept very clean at all times by washing thoroughly in hot water with liquid soap. Efficient drying of hands, which is also important, may be achieved by using disposable paper towels, hot air dryers or continuous roller towels.

Food handlers must wash their hands regularly throughout the working day and especially:

(1) after visiting the w.c.;
(2) on entering the food room and before handling any food or equipment;
(3) in between handling raw and cooked food;
(4) after combing the hair;
(5) after eating, smoking or blowing the nose;
(6) after handling waste food or refuse.

As fingernails may harbour bacteria, they must be kept short and clean. Nail varnish may contaminate food and should not be used. People who continually put their fingers in their mouth, for example, nail biters, should not be employed as food handlers. Licking the fingers before picking up sheets of wrapping paper is a particularly bad habit.

The nose, mouth and ears

Up to forty per cent of adults carry staphylococci in the nose and mouth. Coughs and sneezes can carry droplet infection for a considerable distance and persons with bad colds should not handle open food. Disposable single-use paper tissues are preferable to handkerchiefs. Picking or scratching the nose is not acceptable.

As the mouth is likely to harbour staphylococci, food handlers should not eat sweets, chew gum, taste food with the finger or an unwashed spoon or blow into glasses to polish them. Apart from being aesthetically unacceptable, spitting can obviously result in food contamination and is illegal.

Discharges from the ears, eyes and nose may contaminate food and employees must report these ailments to their supervisor. Medical clearance to start work will normally be required.

Cuts, boils, whitlows and septic spots

Cuts, spots and sores provide an ideal place for bacterial multiplication. To prevent contamination of food by harmful bacteria and blood these lesions should be completely covered by waterproof dressings, preferably coloured blue or green to aid detection if they became detached. Cuts on fingers may need the extra protection of waterproof fingerstalls. Waterproof dressings will also assist in preventing cuts going septic.

Food handlers with colds, boils and septic cuts must not handle food.

Jewellery and perfume

Food handlers should not wear earrings, watches, jewelled rings or brooches, as they harbour dirt and bacteria. Furthermore, stones and small pieces of metal may end up in the food and result in a customer complaint.

Strong-smelling perfume should not be worn by food handlers, as it may taint foods, especially those with a high fat content.

The hair

Hair is constantly falling out and, along with dandruff, can result in contamination of food. Furthermore, the scalp often contains harmful bacteria and must be shampooed regularly. Foodhandlers should wear suitable head covering which completely encloses the hair. Hairnets worn under turbans, helmets and hats are recommended. Combing of hair and adjustments to head covering should only take place in cloakrooms and should not be carried out whilst wearing protective clothing, as hairs may end up on the shoulders and then in the product.

Smoking

It is illegal to use snuff and tobacco, including cigarettes, pipes or cigars, in food rooms or whilst handling open food. Not only is this to prevent cigarette ends and ash contaminating food but also because:

(1) people touch their lips whilst smoking and they may transfer harmful bacteria to food;

(2) smoking encourages coughing and droplet infection;
(3) cigarette ends contaminated with saliva are placed on working surfaces;
(4) an unpleasant environment may be created for non-smokers.

Protective clothing

All food handlers should wear clean, washable, light-coloured protective clothing, preferably without external pockets. Protective garments should be appropriate for the work being carried out and should completely cover ordinary clothing. Jumper and shirt sleeves must not protrude and, if short-sleeved overalls are worn, only clean fore-arms must be visible.

Staff must be aware that protective clothing is worn to protect the food from risk of contamination and not to keep their own clothes clean. Dust, pet hairs and woollen fibres are just a few of the contaminants carried on ordinary clothing.

Outdoor clothing and personal effects must not be brought into food rooms, unless stored in suitable lockers.

General health and reporting of illness

Food handlers should be in good health in all aspects from oral hygiene to general fitness. Any food handlers suffering from diarrhoea, vomiting or suffering from a food-borne infection must not handle food. They must notify their superviser who must contact the local health authority so that environmental health officers can carry out investigations. Food handlers who have consumed a meal known to have caused food poisoning or live in the same household as a confirmed case or have suffered from diarrhoea or vomiting whilst abroad should also report to the superviser. Food handlers who excrete food poisoning organisms must not resume food handling duties without medical clearance.

Persons with skin infections, sores, heavy colds and ear or eye discharge should be excluded, until medical clearance has been obtained.

Hygiene education

All food handlers must receive the appropriate food hygiene education to ensure that they are aware of the dangers of poor food hygiene and so that they have the knowledge to break the chain of events resulting in outbreaks of food poisoning.

The storage and temperature control of food

Correct storage of food is fundamental to the hygienic operation of any food business. Failure to ensure satisfactory conditions of cleanliness, temperature, humidity and stock rotation can result in problems of unfit or spoiled food, including mould, discolouration, staleness and insect and rodent infestations.

Storage conditions should ensure that the nutritional value, appearance, taste and fitness of food are of the highest standard. Storage areas must not be overloaded and the available space must be taken into account when purchasing food.

Dry-food stores

Rooms used for the storage of dried and canned foods should be dry, cool, well-lit, ventilated, vermin-proof and kept clean and tidy. Food should be stored off the floor on suitable shelves such as tubular stainless steel racks, or in mobile bins. Spillages should be cleared away promptly. All goods should be checked on arrival before placing in storage. Problems encountered include soiled delivery trays, infestations, damaged and leaking cartons, rusty cans and out-of-date stock. If possible, fruit and vegetables should be stored in dry, cool, well-ventilated areas, preferably separate from other food. Tubular or mesh shelves are preferred. Fruit should be examined regularly as mould spreads rapidly.

Canned foods

The risk from canned foods is very small compared with the number produced and this safety record will continue if:

(1) blown cans are not used;
(2) badly-dented, seam-damaged, holed or rusty cans are rejected;
(3) stock rotation is carried out.

Blown cans must not be used.

The shelf-life of canned foods

	Months
rhubarb, pasteurized solid meat packs (refrigerated)	9
fruit juice, prunes, milk products	12
new potatoes, blackberries, raspberries, plums	18
vegetables, baked beans, soups, ready-meals	24
solid-pack cold meat products and fish in oil	60

After the above times the food will not present a health risk but there may be changes in colour, texture and flavour.

The storage of perishable food

High-risk perishable foods may be contaminated by harmful bacteria which can multiply to dangerous levels if not stored under refrigeration.

The correct use of refrigerators

The common food poisoning organisms are incapable of multiplying and producing poisons at temperatures below 5°C. Furthermore, the spoilage of food by bacteria and mould is reduced. As can be seen from the "ten commonest reasons for food poisoning" (page 13), temperature control is the single most important factor in preventing food poisoning. Therefore, it is essential for food handlers to receive clear instructions on the use of refrigerators to ensure that they operate effectively.

Siting

Refrigerators should be sited in well-ventilated areas away from heat sources and the rays of the sun.

Construction

Refrigerators should be constructed to facilitate easy cleaning. Internal linings and shelves should be impervious and non-corroding. Door seals must be maintained in good condition and the unit should be serviced regularly.

Operating temperature

Units should normally operate between 1°C and 4°C. A thermometer should be permanently positioned in the warmest part of the refrigerator and the temperature checked at least daily.

Defrosting and cleaning

Defrosting and cleaning should be carried out frequently in accordance with the manufacturer's instructions. Units which defrost automatically should still be cleaned at least weekly. Bicarbonate of soda (1 tablespoon to a gallon of water) may be used, but perfumed cleaning agents must not.

Packing and stock rotation

Refrigerators must not be overloaded, and food should never be placed in front of cooling units. Only perishable foods should be stored in the refrigerator. This includes vacuum packs and pasteurized cans of meat. Stock rotation is essential to avoid spoilage.

Refrigerators must not be overloaded.

Hot food

Hot food must never be placed directly into a refrigerator as this will raise the temperature of food already being stored. It will also encourage condensation and consequently contamination. Joints of meat should, if possible, be kept below $2\frac{1}{2}$ Kgs (6lbs), to facilitate cooling within $1\frac{1}{2}$ hours prior to refrigeration.

Contamination and covering of food

Raw food must always be kept apart from high-risk food. Separate refrigerators are preferred, although, if in the same unit, the raw food must always be placed at the bottom. Food should be covered to prevent drying out, cross-contamination and absorption of odour.

Open cans of food

To avoid the acid attack on opened and part-used cans, they should not be stored in refrigerators, especially such food as fruit, fruit juice or tomatoes. The unused contents should be emptied into a suitable container, such as a covered plastic bowl.

Staff training and responsibilities

All food handlers must receive training on the correct use of the refrigerator, and in particular on keeping the door open for the minimum amount of time. The temperature of refrigerated deliveries should be checked on arrival.

The storage of frozen food

The temperature of deliveries of frozen food should be checked before unloading. If the temperature is around −18°C the food should be transferred to the freezer as quickly as possible. Deliveries above −10°C are unacceptable. New stock should always be placed below existing food. Food must not be stored above the freezer load line and must not be kept for longer than recommended by the manufacturer, as the quality gradually deteriorates.

A general guide for food kept at −18°C is:

	Months
vegetables, fruit, most meat	up to 12
pork, sausages, offal, fish, butter and soft cheese	up to 6

The storage time for food stored in the frozen food storage compartment of a refrigerator:

*	not above −6°C	(one week)
**	not above −12°C	(one month)
***	not above −18°C	(three months)
* ***	−18°C to −25°C	(three months or longer, and is capable of freezing fresh food)

Thawing of frozen food

Most food taken from the freezer can be cooked immediately, but poultry and large joints must be completely thawed before cooking. The manufacturer's instructions should always be followed. Thawing of raw meat/poultry must take place in an area entirely separate from other foods which may be exposed to risk of contamination from thawed liquid. This area must never be used for cooked food which is cooling prior to refrigeration. Thawing is best carried out at 10 to 15°C, or in a thawing cabinet. Thawing times in refrigerators vary considerably depending on the temperature, for example, a 1.1 Kg bird takes around 70 hours at 1°C, 40 hours at 5°C, and 13 hours at 10°C to reach 0°C.

Frozen food must not be thawed near high-risk foods.

Rules for handling frozen poultry

(1) keep separate from other foods;
(2) thaw completely in a cool room. Poultry will be ready for cooking when the body is pliable, the legs are flexible and the body cavity is free from ice crystals;
(3) remove giblets;
(4) once thawed, keep in the refrigerator and cook within twenty-four hours;
(5) cook thoroughly and cook the stuffing separately;
(6) all utensils and surfaces used for the preparation of raw meat and poultry should be thoroughly cleaned and disinfected before being used for high-risk food;
(7) eat straight after cooking or, if the bird is to be carved cold, cool it quickly and store in the refrigerator. As with all meats **refrigerated storage is essential within one and a half hours**;
(8) avoid handling the cooked bird.

THAWING AND COOKING TIMES OF FROZEN POULTRY

Oven ready weight Kg (lbs)	Approx. thawing time at room temp. hours	Min. cooking time at 180°C/350°F Gas 4. (in foil) hours
2.25 (5)	15	$2\frac{1}{2}$
4.5 (10)	18	$3\frac{1}{2}$
6.75 (15)	24	$4\frac{3}{4}$
9.0 (20)	30	$5\frac{3}{4}$

These are minimum times. The bird is cooked when the juices run clear.

Stock rotation

Satisfactory rotation of stock, to ensure that older food is used first, is essential to avoid spoilage. Stock rotation applies to all types of food. Daily checks should be made on short-life perishable food stored in refrigerators, whereas weekly examination of other foods may suffice.

Stock which is undisturbed for long periods will encourage rodent and insect infestations. Good stock rotation has the added advantage of assisting in the maintenance of the correct levels of stock.

Stock rotation has been much easier since the advent of open-date coding but some products do not require a "sell-by" date and, in these cases, food handlers should adopt their own code to identify the date of delivery. Remember the rule: **"First in, first out".**

Preservation

Several methods of preservation are used to prolong the storage life of food by preventing bacteria from multiplying, for example, canning, drying, freezing, heat treatment, vacuum packing and the use of salt or sugar.

The preparation, cooking and cooling of food

The observance of good hygiene practices during food preparation is an important element in preventing food poisoning. Raw food and high-risk food should be prepared in separate areas with separate equipment. The handling of food should be minimized and it must not be left in warm humid atmospheres. Food handlers should work in a logical, planned manner, ensuring that working surfaces are kept as tidy as possible. Spillages and waste food should be cleared away promptly.

Thorough cooking is important to destroy harmful bacteria, although some will produce poisons which will withstand boiling for at least thirty minutes. Spores may also survive cooking.

After cooking, the food should be eaten as soon as possible. If the food is to be kept hot prior to serving, it must be maintained above 63°C.

Food which is to be refrigerated should be cooled quickly and placed in the refrigerator within $1\frac{1}{2}$ hours of cooking. It is better to use smaller joints of meat as they cool more rapidly.

Rewarming of refrigerated meat dishes is bad practice. If food is reheated, it must be cooked thoroughly for immediate consumption.

The time between:

refrigeration and cooking (or processing);
cooking and eating;
cooking and refrigeration;
refrigeration and serving;

must be kept as short as possible.

Food must not be left in the danger zone of temperature.

The construction and design of food premises

The selection of a suitable site is most important when planning food premises. Consideration must be given to the provision and availability of services, i.e. electricity, gas, water supply and effluent disposal and the accessibility for delivery and refuse collection.

After obtaining planning permission, the following design principles should be followed:

(1) cross-contamination should be eliminated; clean and dirty processes must be separated;
(2) work flow should be continuous and progress in a uniform direction from raw material to finished product;
(3) suitable and sufficient facilities for personal hygiene, cleaning and disinfecting equipment and washing food must be provided, together with adequate supplies of hot and cold water;

The only way from a W.C. must be via a wash-hand basin.

(4) adequate refrigeration, cooking and cooling facilities must be provided;
(5) the premises must be capable of being thoroughly cleaned;
(6) insects, rodents and birds must be denied access;
(7) suitable staff facilities must be provided;
(8) adequate drainage capable of removing peak loads quickly without flooding must be provided;
(9) suitable and sufficient ventilation is necessary to ensure reasonable working conditions and reduce temperatures and humidities;
(10) high standards of lighting are necessary to facilitate cleaning and provide safe and satisfactory working conditions.

Construction details

Ceilings

Ceilings should be smooth, fire resistant, durable, light-coloured, coved at wall joints and easy to clean.

Wall finishes
Finishes should be smooth, impervious, non-flaking, durable, light-coloured and capable of being thoroughly cleaned. Surfaces may need to be resistant to spillages, chemicals, grease, heat and impact. Internal walls should be solid, as cavities provide harbourage for pests.

Floor surfaces
Surfaces should be durable, non-absorbent, anti-slip, without crevices and capable of being effectively cleaned. They may need to be resistant to acids, grease and salts, and should slope sufficiently for liquids to drain to trapped gulleys. The angle between walls and floors should be coved.

Windows
Windows, if present, should be fixed on north-facing walls to reduce glare and solar heat gain. Opening windows must be fitted with cleansable fly-screens.

Wooden finishes
Wood should not normally be used but if unavoidable, for example, window frames, it should be well-seasoned, properly knotted, stopped and primed and given three coats of polyurethane paint.

The storage and disposal of waste
Suitable receptacles should be provided, both inside and outside food premises, for the disposal of waste food and debris. Disposable polythene sacks or plastic bins are usually provided for internal use and dustbins or skips for external use. Dustbins must have tight-fitting lids. Compactors which completely enclose the refuse are preferred.

Refuse containers used internally must be emptied as frequently as necessary and always at the end of the day. After emptying, re-usable containers must be thoroughly cleaned before being brought back into the food room. Waste-disposal units are commonly used for removing food debris.

Receptacles used for storage or collection of refuse should not be re-used for the storage of food.

Empty food containers which are intended for re-use, for example, bottles and trays, must be protected from contamination, especially by dogs. It is preferable to store such containers in a suitable room or well clear of the floor under a lean-to.

Keep refuse areas tidy.

External refuse areas must be kept clean and tidy so as not to attract rodents, birds and insects. Receptacles must be emptied as frequently as necessary and the yard surface must be hosed down regularly to avoid creating a nuisance. After emptying, waste receptacles should be rinsed out.

Equipment for food handling

All equipment, working surfaces and other utensils which are used in food premises should be designed and constructed to minimize harbourage of soils, bacteria or pests, and to enable them to be thoroughly cleaned and disinfected. Surfaces in contact with food should be smooth, impervious, non-toxic, non-flaking, corrosion-resistant, durable and suitable for their intended use. Resistance to heat and attack by acid foods, such as fruit juices and milk, may need to be considered. Cracked, chipped, broken and badly-pitted equipment harbours dirt and bacteria and should not be used.

Food-grade stainless steel is appropriate for most equipment. Wooden surfaces should not be used, as they are absorbent and incapable of being cleaned and disinfected. Cutting boards and handles of knives and brushes may be made from polypropylene or other suitable synthetic material. Tubular stainless steel is preferred to painted or galvanized angle iron.

The use of different colours or shapes as a code to ensure equipment used for raw food is not used for high-risk food is recommended.

Where practicable, and with due regard to safety, equipment should be mobile to facilitate its removal for cleaning. This is particularly important if sited close to walls.

Wooden, dirty and defective equipment must not be used.

Cleaning and disinfection

Soiling of surfaces and equipment is unavoidable in all food businesses. It is essential that such residues are not allowed to accumulate to levels which expose food to risk of contamination. Removal of food residues, dirt and grease is the process of cleaning.

The reasons for cleaning

(1) to ensure a pleasant and safe working environment;
(2) to promote a favourable image to customers;
(3) to remove matter on which bacteria would grow, thus reducing the risk of food poisoning and spoilage;
(4) to allow the disinfection* of specific equipment and surfaces;
(5) to remove materials which would encourage pest infestations;
(6) to reduce the risk of foreign matter contamination.

Energy in cleaning

Cleaning is the application of energy to a surface, with the intention of removing dirt and grease.

Energy is applied as:

physical, for example, scrubbing;
heat, for example, hot water;
chemical, for example, detergent.*

Improving chemical energy and increasing heat will reduce the amount of physical energy required.

After cleaning, disinfectants* may be used to destroy bacteria that remain. Hot water, around 82°C, steam and bleach are the commonest disinfectants.

**Detergent – a chemical used to remove grease, dirt and food particles.*

**Disinfection – the reduction of micro-organisms to a level that is safe and which will not cause premature food spoilage.*

**Disinfectant – a chemical used for disinfection.*

Hot water, chemicals and physical energy must be used to clean.

Effective cleaning

To be effective, cleaning must be planned. A schedule which stipulates the frequency, method of cleaning, the amount and type of chemical to use and the person responsible, must be drawn up and implemented.

Staff must be trained to "**clean as they go**" and they must always have regard to the provision of health and safety. Suitable protective clothing must be worn and the chemical manufacturer's instructions must always be followed. Some chemicals can be very dangerous if mixed. Open food must not be exposed to risk of contamination during cleaning.

The cleaning procedure

Cleaning and disinfection normally consists of six basic stages:

(1) pre-clean: removing excess soil by sweeping, wiping or pre-rinsing;
(2) main clean: loosening of the surface grease and dirt using a detergent;
(3) rinse: removal of loose dirt and detergent.
(4) disinfection: destroying micro-organisms using, for example, bleach;
(5) final rinse: removal of disinfectant;
(6) drying: preferably natural by evaporating dry.

In light-soil conditions the pre-clean may be combined with the main clean. If air drying is not possible, single-use paper towels or a clean, dry cloth should be used.

Double-sink washing

Although the use of mechanical dishwashers and glasswashers is becoming more common, double-sink washing is recommended when suitable dishwashing machines are not available. The following procedure should be followed:

(1) remove any heavy or loose soil by scraping and rinsing in cold water;
(2) place articles in the first sink in detergent solution at 50°C to 60°C, scrub with a nylon brush and/or wipe with a clean cloth to loosen dirt residues. Rubber gloves will be required. Cool or dirty water should be replaced;
(3) re-immerse in the first sink to wash off loosened dirt;
(4) place articles in the second sink to rinse off chemical residues;
(5) leave for 30 seconds at 82°C to achieve disinfection;
(6) remove the articles, allow to drain and air dry on a clean, disinfected surface. After drying, store in a clean place free from contamination.

Pest control

A **food pest** is an animal which lives in or on man's food and is destructive, noxious or troublesome. The common pests found in the food industry include:

(1) rodents: rats and mice;
(2) insects: flies, wasps, cockroaches, psocids, silverfish, stored product insects and ants;
(3) birds: mainly feral pigeons and sparrows.

Regular surveys of food premises must be carried out to ensure that they are pest free. Signs to look for include:

(1) live or dead bodies, including larvae and pupae;
(2) droppings;
(3) damage including gnawing marks, holes in sacks, boxes and packets;
(4) spillages adjacent to sacks of food;
(5) rodent smears around pipes.

Reasons for control:

(1) to prevent the spread of disease;
(2) to prevent wastage of food;
(3) to prevent damage;
(4) to comply with the law.

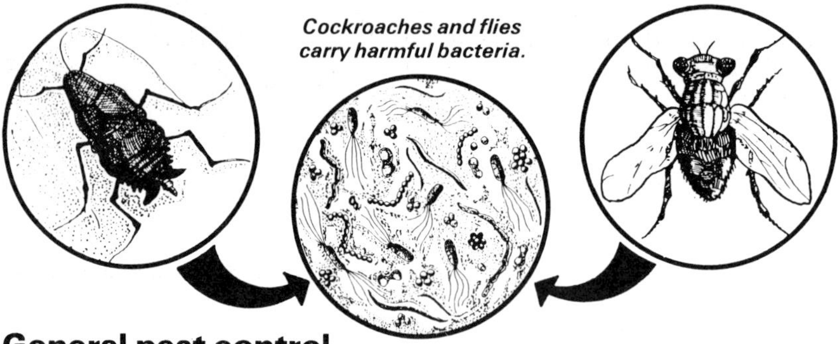

Cockroaches and flies carry harmful bacteria.

General pest control

Pests require food, shelter and security. Denial of these environmental factors will prevent their survival, and is the first line of defence against possible infestations. Environmental control may be considered as denial of:

(1) access – by care in the design, maintenance and proofing of buildings;
(2) food and harbourage – by good housekeeping.

Good housekeeping

To reduce the risk of infestation, it is important to deny the lone invader the conditions it likes, and in particular to ensure that:

(1) premises and refuse areas are kept in a clean and tidy condition;
(2) spillages are cleared away promptly;
(3) food is stored off the floor and clear of walls to facilitate regular inspection;
(4) food is kept in rodent-proof containers and lids are always replaced;
(5) all deliveries of raw materials are checked to ensure their freedom from infestation;
(6) vegetation and other cover in the immediate vicinity of the food premises is removed.

Physical and chemical control

Although very important, environmental control may not be entirely successful and eradication must be carried out using physical or chemical control methods.

Physical control methods are usually preferred as the pest is caught, either dead or alive, and consequently is not able to die in food or some inaccessible place. Examples of physical control include electric fly killers, rodent traps and mist nets for birds. Unfortunately, physical methods are not always successful and poisons have to be used. Rodenticides are used to kill rats and mice and insecticides to kill insects.

Infestations of food premises should be dealt with immediately and food handlers should seek assistance from a local authority or specialist contractor.

Rodents must be controlled.

Essential points to remember about the prevention of food poisoning

(1) To prevent food poisoning **break the chain.**
(2) Raw food and cooked food must be kept entirely separate.
(3) Keep food out of the **danger zone of temperature**.
(4) Wash hands frequently and especially after using the w.c. or handling raw meat.
(5) Persons with diarrhoea, vomiting or septic cuts and boils must not handle food.

The law relating to food and food hygiene

The law is a complex subject and most acts and regulations affecting the food industry are difficult to comprehend. However, ignorance of the law is no defence in the event of a prosecution, and all food handlers should make special efforts to understand the legislation that affects their business and themselves. This book contains only very brief details of some of this legislation and more information may be obtained from the local environmental officer.

Acts and regulations applicable to the food industry are concerned with:

(1) the production or sale of injurious, unsound or unfit food;
(2) the contamination of food;
(3) the hygiene of food premises, equipment and personnel;
(4) hygiene practices, including temperature control and heat treatment;
(5) the control of food poisoning and food-borne diseases;
(6) the composition and labelling of food.

Food Act, 1984

This is the most important Act relating to the sale of food for human consumption and also the Act under which all hygiene regulations are made.

Under this Act it is an offence to sell food which is unfit or not of the nature or substance or quality demanded by the purchaser. Food which is rancid, slimy, mouldy, decomposing or contains poisonous substances or foreign bodies, may therefore be dealt with by the environmental health officer.

Local authorities can apply to the court for a Closure Order to close premises or stalls which are insanitary and pose an immediate threat to public health.

The Food Hygiene (General) Regulations, 1970

The standards of hygiene within food premises are normally controlled by these Regulations although some premises such as dairies and slaughterhouses have their own specific legislation. The main objective of the Regulations is to prevent outbreaks of food poisoning.

The Regulations may be divided into the following sections:

(1) *premises* must be kept in good repair, clean, and proofed against the entry of pests. Food rooms must have satisfactory standards of

lighting and ventilation and refuse and filth must not be allowed to accumulate in these rooms;

(2) *equipment* must be made of non-absorbent material, maintained in good condition, and kept clean. This also applies to food containers;

(3) *food handlers* must keep clean, cover cuts with waterproof dressings, wear suitable protective overclothing, must not smoke or spit and must report to the person carrying on the food business if they are suffering from food poisoning or a food-borne disease. (The employer must inform the local authority);

(4) *washing facilities* – wash-hand basins for securing personal cleanliness, and sinks for washing food or equipment must be provided and kept in a clean condition. Soap, nailbrush and drying facilities must be provided at the wash-hand basin. Hot and cold water or hot water at a suitably controlled temperature must be supplied to both types of unit;

(5) *services* – a satisfactory supply of water, first-aid materials and accommodation for outdoor clothing must be provided. Sanitary conveniences must be well-lighted and ventilated and kept clean;

(6) *practices* include restrictions on exposing food to risk of contamination. In addition, high-risk foods must be kept out of the **danger zone of temperature** (see page 8). Open food exposed for sale must be covered;

(7) *penalties* – offences are punishable, on conviction, by a fine of up to £2,000 for each offence. In serious cases a sentance of up to two years imprisonment and unlimited fines may be imposed.

The Health and Safety at Work etc. Act, 1974

This is the principal Act concerned with protecting the health and safety of employees and members of the public.

Every employer has a legal obligation to ensure, as far as is reasonably practicable, the health, safety and welfare of employees. Safe systems of work must be devised, and supervision and training given to all staff, for example, before staff are allowed to operate or clean dangerous machines, such as gravity feed slicers.

Employees must ensure that their acts or omissions do not adversely affect other persons.

Environmental health officers

Environmental health officers are experienced, highly-trained officers employed by local authorities to ensure the maintenance of high standards of hygiene and the production of safe wholesome food. Although they are responsible for enforcing food legislation, they are always available to give guidance and advice.

APPENDIX I

General rules for food handlers

(1) Food handlers with skin, nose, throat or bowel trouble must inform their supervisor and must not handle food until medical clearance has been obtained.

(2) Cuts, burns and sores must be covered with waterproof dressings. Persons with boils or septic cuts must not handle food. Fingernails must be kept clean and short, and nail varnish should not be used.

(3) Suitable protective clothing and head covering must be worn by all food handlers. Outdoor clothing must not be brought into food rooms. Jewellery, hair grips and watches should be removed.

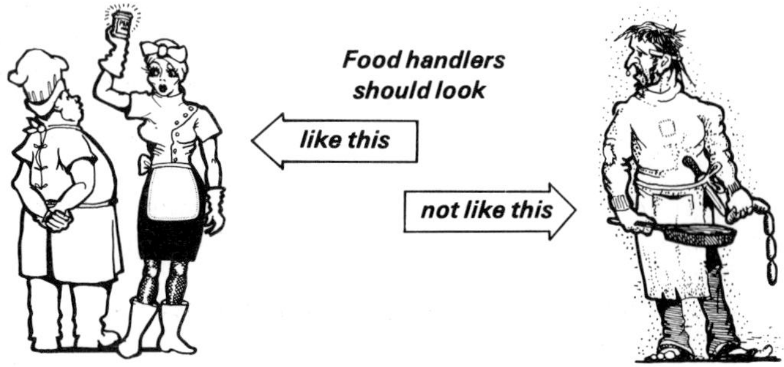

(4) Food handlers must not smoke in food rooms or whilst handling open food.

(5) On entering a food room all food handlers should wash their hands. Hands must also be thoroughly washed after visiting the w.c., handling raw food, blowing the nose, handling refuse or swill, eating, smoking, cleaning and at frequent intervals during the work period.

Food handlers must not pick their nose.

(6) The mouth, nose or hair must not be touched during food preparation. Sneezing or coughing over food must be avoided.

(7) All equipment, fixtures and fittings must be clean before preparation begins.

(8) Raw food must always be kept separate from high-risk food at all stages of storage and preparation. Separate equipment, working surfaces and food handlers should be used to avoid cross-contamination from raw to high-risk food.

(9) Frozen meat must be completely thawed before cooking. Thawing must be carried out separate from other processes. The liquid from defrosting poultry, etc., must not come into contact with other food, directly or indirectly, for example, by use of a wiping cloth.

(10) All meat must be cooked thoroughly. Rewarming cooked meats is potentially hazardous. All reheated foods including meat, soup and gravy must be thoroughly reheated and consumed immediately.

(11) If hot meat is not to be eaten immediately, it must be cooled quickly and placed in a refrigerator within $1\frac{1}{2}$ hours. Cooked food must be protected from contamination at all times and kept out of the **danger zone of temperature,** i.e. 10°C to 63°C.

(12) Food must not be removed from the refrigerator until required for serving or preparation.

(13) Display trolleys and buffets should be refrigerated and covered to prevent contamination.

(14) Stored food must be rotated to prevent spoilage, avoid waste and prevent infestations. Out-of-date food and damaged canned or packaged food must not be sold.

(15) Dirty pans, cutlery, crockery and other equipment should be cleaned and sterilized without undue delay.

(16) If the use of drying cloths is unavoidable, they must be kept clean.

(17) Dirty wiping cloths must not be used. Disposable cloths are preferred.

(18) Spillages should be cleaned up promptly and food debris should not be allowed to accumulate.

(19) Polythene sacks or suitable impervious containers, fitted with lids, should be used for the disposal of waste in food rooms.

(20) Animals must not be allowed in food rooms. Infestations of insects or rodents must be reported immediately.

REMEMBER,
YOUR CARELESSNESS
MIGHT BE THE CAUSE
OF AN OUTBREAK OF
FOOD POISONING

APPENDIX II

The hygienic preparation and display of cooked meat

(1) All cooked meat and pasteurized canned meats should be stored at or below 4 °C. The preparation period, when the goods are not under refrigeration, should be as short as possible.

(2) Staff suffering from colds, diarrhoea or vomiting, septic cuts and sore throats, must not handle meats and should notify their supervisor immediately.

(3) The food preparation surfaces, the can-opener and the slicing machine must be thoroughly cleaned with a detergent and a disinfectant and rinsed immediately prior to their use in connection with the preparation or slicing of cooked meats.

Separate equipment should be used for bacon slicing and cooked meats. Raw meats must not be allowed to come into contact with cooked meats or surfaces used for cooked meat. Knives and other equipment used for raw meat must never be used for cooked meat.

(4) Food handlers must wash their hands immediately prior to handling cooked meat and frequently throughout the day. Handling of the meats by staff should be kept to the absolute minimum. Suitable protective overclothing and head covering must be worn. Staff must **never** handle cooked meat after handling raw meat, unless they have thoroughly washed their hands.

(5) All cooked meat should be displayed at or below 4 °C. It must **not** be displayed for sale at ambient temperatures.

(6) All products on display must be examined daily to ascertain that they are satisfactory and are within their shelf-life. The manufacturer's advice must be adhered to.

Food handlers who are ill must not handle food.

APPENDIX III

The protection of food in retail premises

Sources of contamination	Vehicles/ routes of contamination	Prevention of contamination
Customer	Coughing/or sneezing	Cover food, provide screens, remove accessibility of the food; customers should not be able to handle open food. Pre-wrap
	Handbags and shopping bags on counter	Provide rail. Counter should not be used for food storage or preparation
	Smoking	Display 'no smoking' notices
	Animals	Keep animals out of food premises. Store food well clear of the floor

Food must be protected.

Staff	Poor personal hygiene	Training. Ensure high standards maintained by management. Provide wash-hand basins that are easily accessible and close to work stations
	Blowing into bag Licking fingers Picking nose Eating sweets	Regular training sessions and effective supervision
	Cuts or sores	Provide waterproof dressings

Sources of contamination	Vehicles/ routes of contamination	Prevention of contamination
	Sneezing or coughing	Move away from food, and wash hands after using handkerchief. Exclude persons with colds
	Handling open food	Use clean tongs Effective training and supervision
	Dirty overclothing	Head covering and suitable protective clothing should be provided
	Smoking	Prohibited
Dirty premises or equipment		Provide suitable cleaning materials and effective cleaning schedules (including sufficient time and staff)
Stale, mouldy or unfit food		Effective stock rotation Effective storage facilities (normally includes correctly maintained refrigeration units)
Rats, mice and insects		Good housekeeping Effective proofing Efficient pest control

KEEP IT CLEAN – KEEP IT COOL – KEEP IT COVERED

Smoking in food rooms is illegal.

APPENDIX IV

Safety in the kitchen

It is essential that all persons engaged in the preparation, cooking and service of food are aware at all times of the need for certain elementary precautions of safety, in order to automatically minimize the risk of accident.

(1) Do not leave metal spoons in boiling liquids.
(2) Do not leave handles of cooking pans over the gas flame.
(3) Always carry a knife with its point towards the floor.
(4) Never attempt to catch a falling knife.
(5) Always cut or chop on a board. Never in the hand.
(6) When using a hand model can-opener, protect the other hand with a cloth or kitchen towel.
(7) Always clean up any spilled grease or fat from the floor immediately.
(8) Never use a damp cloth for lifting or carrying hot utensils.
(9) Never wear sandals or open shoes in the kitchen.
(10) Pans containing hot fat which catch fire should be extinguished by smothering with a fire blanket or a thick damp sack. Never pour water on a fat or oil fire.
(11) Long hair styles should be 'put up' and covered with a clean head cap or similar means of protection.
(12) For economy, as well as safety, all gas and electrical appliances must be turned **off** when not in use.
(13) Make sure that the **first aid** box is readily accessible and that it contains sufficient waterproof dressings and burn dressings.
(14) Never attempt to carry large heavy containers of hot food single-handed. Get assistance.
(15) Do not reach over naked burning gas appliances.
(16) Always ensure that gravity meat slicers are adequately guarded.
(17) Ensure that all drainage channel covers are properly fitted.
(18) Never put cleaning fluids into bottles originally used for food or drink e.g. lemonade or milk bottles.

By courtesy of the City of Canterbury Environmental Health Department.

Definitions

Antibiotic	A drug used to destroy pathogenic bacteria within human or animal bodies.
Antiseptic	A substance that prevents the growth of bacteria and moulds, specifically on or in the human body.
Bactericide	A substance which destroys bacteria.
Carrier	A person who harbours, and may transmit, pathogenic organisms without showing signs of illness.
Cleaning	The removal of soil, food residues, dirt, grease and other objectionable matter.
Contamination	The occurrence of any objectionable matter in food.
Danger zone of bacterial growth	The temperature range within which multiplication of pathogenic bacteria is possible, (from 10°C to 63°C).
First-aid materials	Suitable and sufficient bandages and dressings, including waterproof dressings and antiseptic. All dressings to be individually wrapped.
Food handling	Any operation in the production, preparation, processing, packaging, storage, transport, distribution and sale of food.
Gastroenteritis	An inflammation of the stomach and intestinal tract that normally results in diarrhoea.
Germicide	An agent used for killing micro-organisms.
Incubation period	The period between infection and the first signs of illness.
Mildew	A type of fungus similar to mould.
Moulds	Microscopic plants (fungi) that may appear as woolly patches on food.
Optimum	Best.
Pathogen	Disease-producing organism.
Pesticide	A chemical used to kill pests.
Residual insecticide	A long-lasting insecticide applied in such a way that it remains active for a considerable period of time.
Sanitizer	A chemical agent used for cleansing and disinfecting surfaces and equipment.
Spores	A resistant resting-phase of bacteria protecting them against adverse conditions, such as high temperatures.
Sterilizer	A chemical used to destroy all living organisms.
Sterile	Free from all living organisms.
Sterilization	A process that destroys all living organisms.
Toxins	Poisons produced by pathogens.
Viruses	Microscopic pathogens that multiply in living cells of their host.
Wholesome food	Sound food, fit for human consumption.